Habitat Days and Nights

DAY AND NIGHT ON THE Tundra

by Mary Boone

PEBBLE
a capstone imprint

Published by Pebble, an imprint of Capstone.
1710 Roe Crest Drive, North Mankato, Minnesota 56003
capstonepub.com

Copyright © 2022 by Pebble, a Capstone imprint.
All rights reserved. No part of this publication may be reproduced in whole or in part, or stored in a retrieval system, or transmitted in any form or by any means, electronic, mechanical, photocopying, recording, or otherwise, without written permission of the publisher.

Library of Congress Cataloging-in-Publication Data is on file with the Library of Congress.
ISBN: 9781663976895 (hardcover)
ISBN: 9781666327717 (paperback)
ISBN: 9781666327724 (ebook PDF)

Summary: Spend a day and night on the tundra! Learn about this cold habitat through the many animals that call it home. Catch breakfast mid-flight with a peregrine falcon. Spend the afternoon snoozing with an Arctic fox. Take an evening trek with a herd of caribou. After dark, sit still with an Arctic hare as it hides from hungry wolves. What will tomorrow bring on the tundra?

Image Credits
AP Images: Dennis Fast/VWPics, 17; iStockphoto: cweimer4, Cover (tundra), 1, dferry, 9, SHIROFOTO, 10, twildlife, 7; Mighty Media, Inc.: 20, 21; Shutterstock: Agami Photo Agency, 12, Alexey Seafarer, Cover (Arctic fox), 1, Dave McKissick, 13, Incredible Arctic, 11, JacobLoyacano, 19, Linda Szeto, 5, RRichard29, 15

Editorial Credits
Jessica Rusick, editor, media researcher; Kelly Doudna, designer, production specialist

All internet sites appearing in back matter were available and accurate when this book was sent to press.

Table of Contents

What Is the Tundra? 4
Morning 6
Noon 8
Late Afternoon 10
Evening 12
Night 14
Late Night 16
Dawn 18
    Tundra Activity 20
    Glossary 22
    Read More 23
    Internet Sites 23
    Index 24
    About the Author 24

Words in **bold** are in the glossary.

What Is the Tundra?

The tundra is a cold **habitat**. It is in the far north and south. The tundra is often snowy and frozen. But plants grow during short summers. There are few trees.

In summer, the tundra sun shines nearly all day. Many animals are active. Some come out during the day. Others come out at night.

The tundra in summer

Morning

The sun rises. A grizzly bear family wanders from its den. The bears **hibernated** for seven months. This helped them survive the cold winter.

The bears did not eat during their long sleep. They are hungry! They search for roots and berries to eat. Grizzly bears also eat plants and fish.

Grizzly bears

Noon

A peregrine falcon soars in the sky. It eats other birds. The falcon spots **prey** from above. Then it quickly dives down. The falcon can dive at 200 miles (322 kilometers) per hour!

Peregrine falcon

Late Afternoon

An Arctic fox hunts in the bright sun. In winter, the fox grows white fur. It blends in with snow. Now, the fox blends in with dirt and rocks. This **camouflage** helps the fox sneak up on prey.

The fox catches a bird. It carries the animal back to its rocky den.

An Arctic fox in winter

An Arctic fox hunting

Evening

A snowy owl hunts in golden light. The owl can see and hear well. It can find prey hiding in grass or snow.

Snowy owl

A lemming in its burrow

The owl hears a lemming in a **burrow**. It silently swoops down. The owl grabs dinner with its **talons**.

Night

Click. Click. A caribou herd is walking. Caribou feet click. The sound comes from **tendons** sliding over foot bones. Clicking helps caribou hear one another.

The herd is walking to a calving ground. Female caribou will have babies here. There are fewer **predators** at the calving ground. There is also more food to eat.

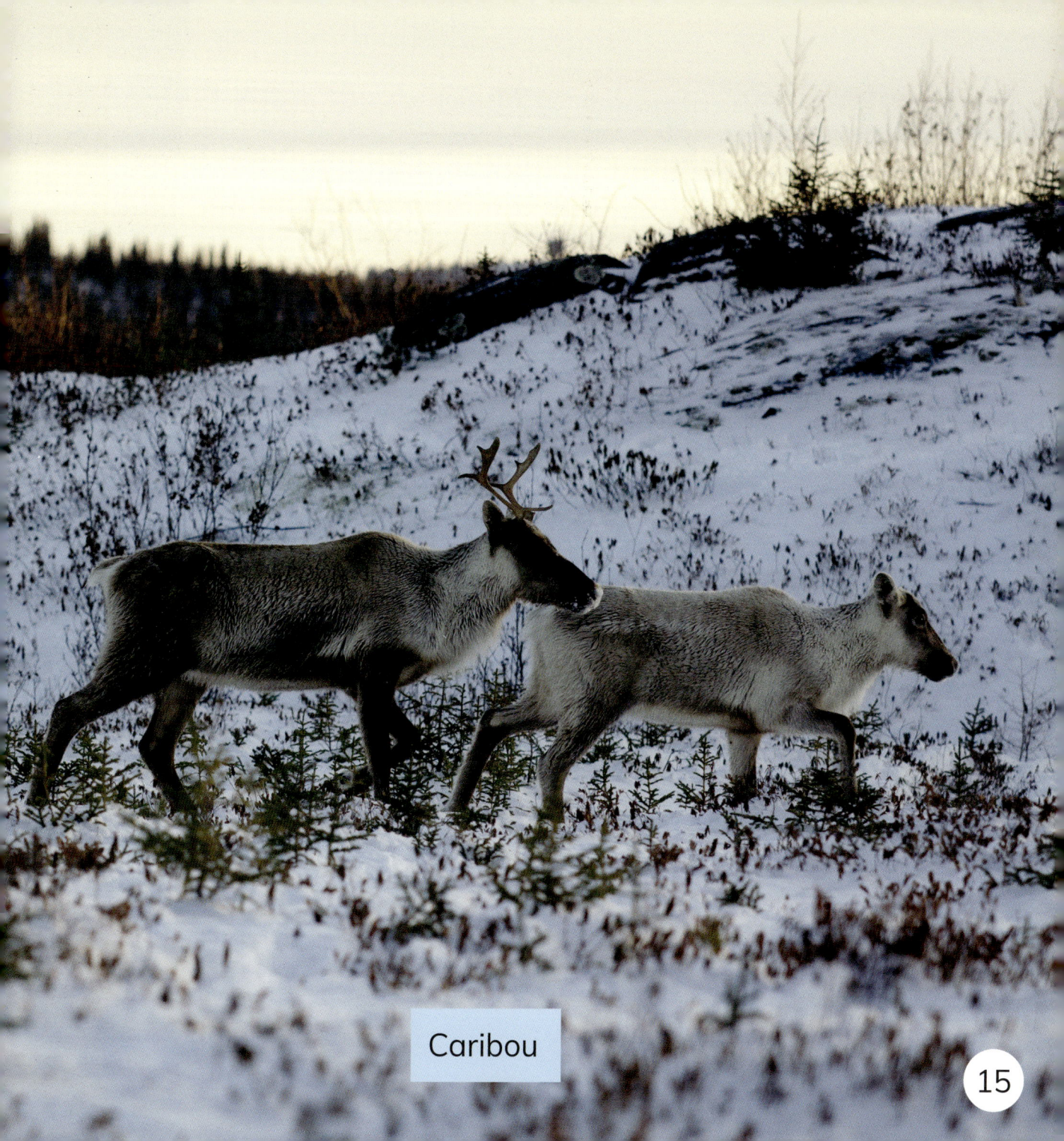
Caribou

Late Night

The sun sits low in the sky. An Arctic hare looks for food. The hare has a strong sense of smell. It finds leaves and berries to eat.

The hare watches for predators as it eats. The hare's eyes are on the sides of its head. It can see almost all the way around itself without moving!

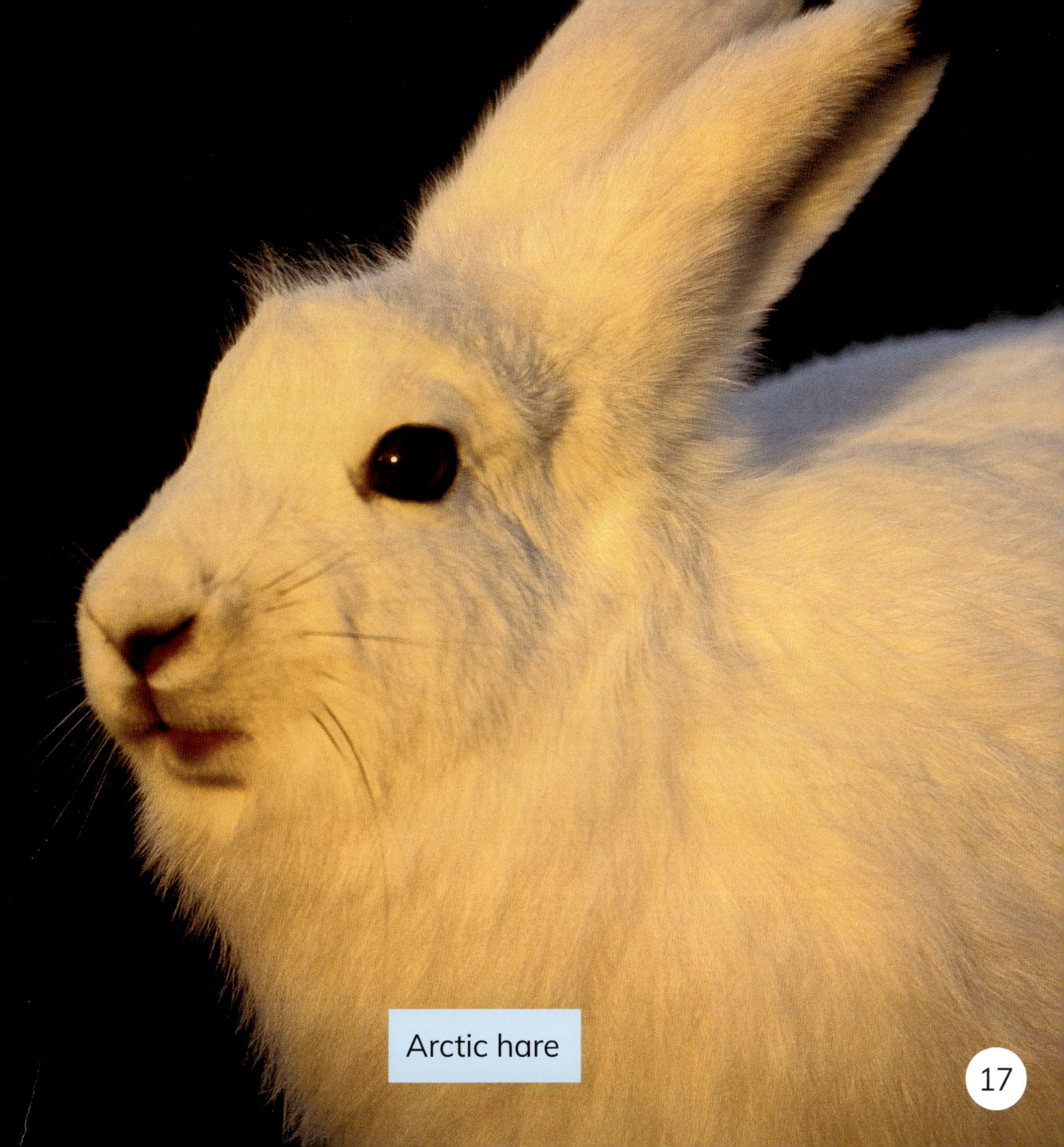
Arctic hare

Dawn

The sun comes up. The air is cool. Musk oxen eat grass. Then the animals rest. As they rest, the oxen spit up their food. They chew it again. This helps their stomachs **digest** the food.

More sleeping animals wake. Another day on the tundra has begun.

Musk oxen

Tundra Activity

What You Need:

- construction paper
- crayons, markers, or colored pencils
- scissors

What You Do:

1. Draw a tundra background on a sheet of blue paper.

2. Draw a vertical line down the center of the paper. Label the left side "Day." Label the right side "Night." Color the "Night" side so it is darker than the "Day" side.

3. Which tundra animals are active during the day? Draw two on another sheet of paper and cut them out. Place them on the "Day" side.

4. Which tundra animals are active at night? Draw two on another sheet of paper and cut them out. Place them on the "Night" side. Then show off your tundra scene!

Glossary

burrow (BUHR-oh)—a tunnel or hole in the ground made or used by an animal

camouflage (KA-muh-flahzh)—a pattern or color on an animal's skin or fur that helps it blend in with the things around it

digest (dye-GEST)—to break down food so it can be used in the body

habitat (HAB-uh-tat)—the natural place and conditions in which a plant or animal lives

hibernate (HYE-bur-nate)—to spend the winter in a deep sleep; animals hibernate to survive low temperatures and lack of food

predator (PRED-uh-tur)—an animal that hunts other animals for food

prey (PRAY)—an animal hunted by another animal for food

talon (TAL-uhn)—a long, sharp claw

tendon (TEN-duhn)—a strong, thick cord of tissue that joins muscle to bone

Read More

Emminizer, Theresa. *Swift Snowy Owls*. New York: PowerKids Press, 2022.

Pfeffer, Wendy. *The Arctic Fox's Journey*. New York: HarperCollins, 2019.

Rustad, Martha E. H. *Animals of the Arctic Tundra*. North Mankato, MN: Capstone Press, 2022.

Internet Sites

Britannica Kids—Tundra
kids.britannica.com/kids/article/tundra/399631

DK Find Out!—Tundra
dkfindout.com/us/animals-and-nature/habitats-and-ecosystems/tundra/

The Kid Should See This—What Are Tundras?
thekidshouldseethis.com/post/what-are-tundras-national-geographic

Index

Arctic foxes, 10
Arctic hares, 16
camouflage, 10
caribou, 14
eating, 6, 8, 14, 16, 18
grizzly bears, 6
hibernation, 6
hunting, 10, 12, 13

lemmings, 13
musk oxen, 18
peregrine falcons, 8
plants, 4, 6, 12, 16, 18
snowy owls, 12, 13

About the Author

Mary Boone has written more than 60 nonfiction books for young readers, ranging from biographies to craft guides. Mary lives in Tacoma, Washington, where she shares an office with an Airedale Terrier named Ruthie Bader.

JAN 25 2022